Halting the Melancholic Tintinnabulation of the Church Tower Bells for Transgender People: Are There Specific Indicators of Suicidality in Transgender & Gender Non-Conforming Individuals?

Halting the Melancholic Tintinnabulation of the Church Tower Bells for Transgender People: Are There Specific Indicators of Suicidality in Transgender & Gender Non-Conforming Individuals?

Matthew R. Lunsford, MS

Matthew R. Lunsford
2017

First Printing: 2017

ISBN: 978-1-365-78553-5

Matthew R. Lunsford
1419 William Street
Fredericksburg, VA 22401

Dedication

To all transgender and gender non-conforming people that are suffering on a daily basis because they are unable to express themselves in a safe way. It is my hope that further research into this population will assist psychologists in preventing further unnecessary deaths in this population.

Contents

Acknowledgements...1

Introduction 2

Chapter I: Initial Resource Collection 3
    Research Question...3
    Summarization of the Literature.....................................3
    Assessment of Sources 4
    Analysis of Sources...8
    Conclusion 13

Chapter II: Literature Review 14
    Research Question 14
How Does Transgender Suicide Fit into the Field of
Psychology?..14
    Synthesis of Research Findings 18
    Evaluation of Studies 19
    Research Gaps 20
    Contribution of Existing Research...............................20
    Conclusion 22

Chapter III: Methodology, Design, and Data.....................24
    Research Method(s) 24
    Support of Method(s)..24
    Strengths and Weaknesses 25
    Data-Collection Tools...25
    Population and Sampling Procedures...........................26
    Data Collection 26
    Data-Analysis Process 28
    Justification of Statistical and/or Qualitative Analysis
....29
    Limitations and Assumptions 29
    Dissemination of Findings...30

Chapter IV: Conclusion 32

Chapter V: Professional Reflection....................................34

Overall Experience and Issues..34

Giving and Receiving Feedback from Peers and Dr.
MacCarty 34

What Led to This Project?......................................34

Taking This Project Into the Workplace and a Doctoral
Program 35

GCC 35

Conclusion 36

Appendix A: List of Acronyms 37

Appendix B: Glossary 39

References 41

Acknowledgements

I would like to acknowledge the contributions of my mom, Dorothy K. "Kay" Lunsford, for being my proofreader and sounding board throughout the process of earning my Master of Science degree and Elizabeth A. "Beth" Streich for serving as emotional support throughout this academic journey.

Introduction

This document examines the phenomenon of suicidality within the Lesbian, Gay, Bisexual, and Transgender community, with a specific focus on transgender individuals. Currently, evidence from research suggests that suicidality is indicated by certain symptoms and behaviors; however, there is currently little to no research into whether or not transgender individuals exhibit unique suicidality markers. The goal of this research is to critically examine the literature surrounding transgender suicide and whether or not there are unique markers that mental health professionals working with transgender patients should be aware of.

Chapter I: Initial Resource Collection

Research Question

The predominant focus of the author's research for this document is transgender suicide. The author hypothesizes that transgender individuals may be susceptible to unique suicidality markers in contrast to the rest of the Lesbian, Gay, Bisexual, and Transgender (LGBT) and/or heterosexual communities. The author wishes to address the specific question of: "Are there unique causal factors in transgender suicide?" This research question would include a H_o of "There are unique causal factors in transgender suicides" and a H_a of "There are not specific causal factors in transgender suicides."

Significance of Transgender Suicide. Currently, psychologists have a fairly vast body of knowledge concerning markers of suicidality, which can be found within the *Diagnostic and Statistical Manual of Mental Disorders-V* (DSM-V), which outlines warning signs to which psychologists should be alert when working with individuals that have certain disorders (American Psychiatric Association [APA], 2013). However, with the fairly recent rise in LGBT people feeling safe enough to explore their sexuality and gender identity, we are seeing a rise in the number of suicides found within the LGBT community, specifically transgender people (Goldblum, et al., 2012). If psychologists wish to reduce the rate of transgender suicides, it is paramount that research be done into the specific causal factors, if any, that surround transgender suicidality.

Summarization of the Literature

The literature obtained so far echoes the majority of the current data that psychologists have on gender identity, sexuality, transgender issues, and suicidality. However, the picture that is emerging is that there is alarmingly high rate of suicide within the transgender community. It is also apparent that suicide as a whole is still a serious issue within clinical psychology; however, research is slowly emerging that is giving further insight into why this is the case. There is currently a vast amount of knowledge concerning why people attempt and/or commit suicide, such as what is found within the DSM-V and the work of Berger (2011).

With that being said, one may argue that not enough research is being conducted within the transgender community and its suicide rates. Goldblum, et al. (2012) suggests that victimization based on one's gender identity may be a large factor into why so many transgender people are attempting and committing suicide; however, the authors also admit that there is a lack of knowledge concerning the markers that go into transgender suicide in addition to a lack of action on the parts of families, school systems, and lawmakers to stop transgender suicide.

From a policy and litigation standpoint, the issue at hand is multifaceted. It can best be thought of as a tiered system; that is, transgender suicide is an issue that affects the transgender individual, the LGBT community, transgendered persons' families, and lawmakers (Goldblum, et al., 2012). At the individual level, transgender people struggle with the psychopathology that surrounds their potential suicidality, whereas the entire LGBT community suffers as well as they are, the majority of the time, not being properly helped by their families and lawmakers that are supposed to be making an effort to ensure that transgender and LGBT individuals are able to receive the appropriate help that they need (Goldblum, et al., 2012; American Psychiatric Association, 2013).

The predominant argument that one may derive from the sources obtained thus far is that suicide is still a serious issue, both in the heterosexual and LGBT communities and that not enough is being done to stop these deaths from occurring due to an apparent lack of knowledge within this subfield. Further analysis will be needed in order to pinpoint specific suicidality markers within the transgender population, if such unique makers are even present within the population.

Assessment of Sources

DSM-V. The DSM-V is a diagnostic manual for mental illnesses that is published by the APA that is used to assist in diagnosing psychopathology (APA, 2013). For the purpose of the author's research question, the DSM-V is useful in that it gives the author an authoritative basis in which to discuss suicidality as well as gender dysphoria (APA, 2013). Rather than presenting arguments or a conclusion, the APA's presentation is one that permits the reader to draw their own conclusions based on the information being presented. That is to say, the DSM-V presents the information regarding the various ways in which psychopathology can manifest itself in such a way that a clinician or researcher can examine the presented information and conduct their own analysis based on what symptomatology or psychological phenomenon are being observed.

The Relationship Between Gender-Based Victimization and Suicide Attempts in Transgender People. Although the article in question does not necessarily deal with suicidality markers, it does address a possible causal factor in transgender suicide (Goldblum, et al., 2012). The article's authors suggest that school-based bullying for one's gender identity can be a casual factor in suicide attempts (Goldblum, et al., 2012). All of the article's authors are either professors of psychology or practicing psychologists, with the exception of Pflum, who is a PhD student (Goldblum, et al., 2012). The authors argue that school-based bullying can in fact lead to transgender individuals attempting and/or committing suicide (Goldblum, et al., 2012). The authors found that out of 290 total respondents, transmen endured the most "gender based victimization (GBV)" at 60.5% in contrast to transwomen's 38.8% "$\chi^2 = 17.4; p = .001$)" (Goldblum, et al., 2012, p. 471).

In terms of suicide attempts, once again the authors found that transmen (32.1%) had a higher rate of suicide attempts than transwomen (26.5%); in addition, those that came from the "low and middle classes had a much higher rate of suicide attempts (30.5% and 29.0%, respectively) in contrast to those from the upper class (9.1%) (Goldblum, et al., 2012, p. 471).

Of additional note, and concern, is the author's findings that GBV caused individuals to "be four times more likely to attempt suicide" in contrast to transgender individuals that did not experience GBV (Goldblum, et al., 2012, p. 471). The figure below outlines these findings in contrast to transmen and transwomen.

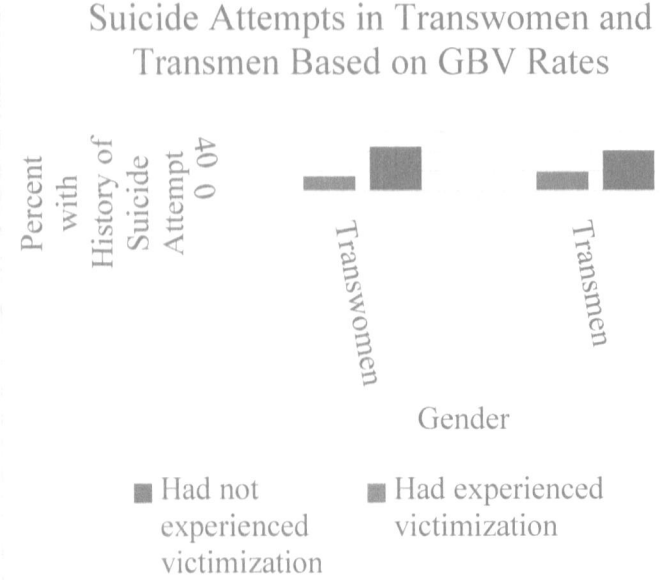

Figure 1: "In-school gender-based victimization and history of suicide attempt" (Goldblum, et al., 2012, p. 472)

Taking the nature of the material into consideration, the authors appear to be highly objective and non-biased in their analysis of the study's subjects. In addition, the study's authors are highly persuasive and note in their findings that the issue of GBV and its suggested connection to suicide attempts in transgender youth is something that "must be addressed at all levels, including the individual, schools, and family," if anything is going to be done to stop the problem in question (Goldblum, et al., 2012, p. 472).

The Developing Person Through the Life Span. *The Developing Person Through the Life Span* is a textbook by Dr. Kathleen S. Berger, who is a professor of psychology at Bronx Community College (Berger, 2011). Berger has taught developmental psychology for the past 20 years, thereby making her an authoritative figure within the field of developmental psychology (Berger, 2011). In terms of informing the author's research, the text provides an excellent analysis of suicidality and gender and sexuality from a developmental standpoint (Berger, 2011). That is, rather than examining the issue of transgender suicide and its causal factors from one standpoint, the author wishes to consider all possible standpoints and possible causation prior to drawing a conclusion into the markers of transgender suicidality.

Berger remains objective and nonbiased throughout the text, predominantly focusing on the data surrounding developmental issues and a general discussion of suicidality and its causal factors that will provide an excellent springboard for the author's research. Berger is highly persuasive throughout her text and concludes that suicidality is predominantly environmentally caused with roots in one's gender identity (Berger, 2011). That is to say, issues such as a pejorative home life or lack of acceptance by friends and family for various reasons, including one's gender identity and/or sexuality, can play a large role in one's choice to attempt and/or commit suicide (Berger, 2011).

Transgender Identity and Suicidality in a Nonclinical Sample: Sexual Orientation, Psychiatric History, and Compulsive Behaviors. Mathy's (2002) article focuses predominantly on the issue of whether or not there is a link between suicidality, suicidal ideation, and one's sexual orientation and gender identity. Mathy (2002) works with the University of Minnesota – Twin Cities' Departments of Psychiatry and Social Work, thus making her an excellent authority on the issues of sexual orientation and being transgender and how these two factors can potentially influence one's choice to commit or attempt suicide. The author finds that this article will be a cornerstone of their research as it provides an excellent foundation to build upon given Mathy's (2002) findings.

It was found that transgender individuals are much more likely to attempt and/or commit suicide than others within the LGBT community (Mathy, 2002). However, the chances of any LGBT individual attempting suicide is magnified if they are currently abusing alcohol or other forms of drugs, but it should be noted that regardless of whether or not drugs and/or alcohol are being utilized, transgender individuals still have the highest chance of attempting and/or committing suicide (Mathy, 2002).

The author does not detect any form of bias within the article; Mathy (2002) does not show any leanings towards preferring LGBT subjects or heterosexual objects and remains objective throughout the text. All data appear to be unmanipulated and presented in a method that facilitates ease of the reader's comprehension.

Analysis of Sources

DSM-V.

Methodology. The DSM-V does not necessarily utilize a specific methodology per se, but rather, as stated above, presents the necessary information in a style that allows for ease of interpretation (APA, 2013). It should also be noted that the DSM-V is put together by a panel of various experts within the field of psychology and psychiatry, thus suggesting that the DSM-V is one the most authoritative sources available (APA, 2013). This "methodology" presents some advantages and disadvantages from an experimental psychological standpoint, which will be discussed in detail below.

Strengths. The predominant strength of the DSM-V is that it is assembled by a panel of experts rather than coming from a myriad of various sources, thus it contains a form of "continuity" (APA, 2013). By continuity, the author means that the DSM-V is self-contained in its authority as a source, thus it may be utilized without a large risk of there being an error in the information that is being presented to the psychologist or researcher.

Weaknesses. Although rather paradoxical, the strength of the DSM-V also appears to be its weakness. As previously stated, the information contained within the DSM-V is assembled by a panel of experts (APA, 2013). Although research is conducted, what goes into each update of the DSM-V is voted on by the experts, so there is little "traditional" peer review to avoid things such as bias (APA, 2013). On that note, the author feels that the DSM-V is a fantastic source, but it should not be used as a "Bible," and should be utilized in conjunction with multiple sources to provide the best picture of the particular psychopathology and/or symptomatology that is being researched.

The Relationship Between Gender-Based Victimization and Suicide Attempts in Transgender People.

Methodology. The authors' methodology involved recruiting participants $(N = 350; n = 290)$ from the Virginia Transgender Health Initiative Survey (THIS) and asking whether or not they had experienced GBV during school and/or their lifetime and whether or not they had a history of suicidality and/or suicide attempts (Goldblum, et al., 2012). The definition of transgender was left intentionally broad to allow for maximum participation and the survey was distributed in English and Spanish; in addition, a financial incentive of $15 was made available to any participant that wished to be compensated for their time (Goldblum, et al., 2012).

The authors' methodological approach allowed for the identification of four categories of transgender people (Goldblum, et al., 2012, p. 471):

- those assigned male at birth and wished to transition to

  women. This is also referred to as a "transwoman" (see

  Appendix B).

- those assigned female at birth and wished to transition to men.

  This is also referred to as a "transman" (See Appendix B).

- those assigned male at birth, but do not identify as fully or

  consistently male and do not identify or wish to transition to

  women

- those assigned female at birth, but do not identify as fully or

  consistently female and do not wish to transition to men. This, and those assigned male at birth, but not wishing to transition,

is often referred to as "non-binary" (see Appendix B).

Of these categories, "participants included 147 transwomen, 81 transmen, 33 nontransitioning trans people that were assigned male a birth, and 29 nontransitioning trans people that were assigned female at birth" (Goldblum, et al., 2012, p. 471). Upon collection of all data, the data were analyzed utilizing standard p-values, chi-squares, and odds ratio calculations (Goldblum, et al., 2012).

In relation to the author's research question Goldblum, et al.'s (2012) article provides an excellent foundation for beginning to understand the causal factors behind transgender suicide attempts. In addition, this article permits the author to have a strong scholarly footing on which to build their analysis of the transgender population, taking into consideration that this is a fairly recent topic in psychological science, thus obtaining sufficient data is somewhat difficult in contrast to more wildly studied topics. Goldblum, el al. (2012) also informs the notion of gender variation within the transgender community. That is to say, the current evidence suggests that if there are unique suicidality markers within the transgender population as a whole, it is possible that there may also be unique factors among the different genders.

Strengths. The strengths of Goldblum, et al. (2012) are numerous. The authors utilized an authoritative source for obtaining access to transgender individuals for which to study. In addition, the authors' methodology appeared to be nonbiased and their suggestions for the social and legal implications of transgender suicide were helpful in guiding this author's research. It should also be noted that Goldblum, et al. (2012) provided an excellent statistical analysis that offered extensive insight into the severity and ramifications of transgender suicide.

Weaknesses. Goldblum, et al. (2012)'s weaknesses were minimal. As stated above, their statistical analysis was excellent and the authors' source was highly authoritative. However, the striking issue here is the lack of participants. Although 290 individuals participated in the authors' research, this is not enough people to offer generalizability to the entire transgender population (Goldblum, et al., 2012). Although the participants were minimal, the article still offered excellent insight into the issue at hand and will guide the author in obtaining additional research.

The Developing Person Through the Life Span.

Methodology. Like the DSM-V, Berger's (2011) text does not utilize a specific "methodology." Berger's text offers an exhaustive examination of the various factors that influence suicide throughout the lifespan. As previously stated, this will offer the author a strong foundation on which to build their analysis of suicidality and how one's gender identity and/or issues surrounding one's gender identity may influence one's choice to attempt or commit suicide.

Strengths. Although Berger's (2011) text is rather general, it offers a thorough analysis of suicide and gender and sexuality in general. In contrast to Goldblum, et al. (2012), Berger (2011) provides an overview of suicide that is not specific to the transgender community. At face value, it would appear that this would harm the author's research; however, this general overview will permit the author to more closely examine the specificity of transgender suicide by analyzing the cases for the general causal factors and markers found within Berger's (2011) text.

Weaknesses. The weaknesses of Berger's (2011) text are minimal. The one issue that the author wishes to note is that this text is basic and is most likely intended for undergraduate students. With that being said, the author is not eliminating the text from consideration as an initial source since the information contained within is highly thorough and will permit the author to analyze scholarly articles that may contain more information regarding suicidality and gender and sexuality. The impact this has on the author's research question is minimal; although the text is basic, it still provides a strong foundation on which to build further scholarly analysis for the author's research project.

Transgender Identity and Suicidality in a Nonclinical Sample: Sexual Orientation, Psychiatric History, and Compulsive Behaviors.

Methodology. Data were obtained via administration of two human sexuality surveys on MSNBC for the month of June, 2000 (Mathy, 2002). The first survey was offered to every "1,000[th] visitor to MSNBC's site and obtained a total of 7,544 responses," or 25% of the visitors to the website (Mathy, 2002, p. 51). The vast majority of the respondents were of minority sexual orientations as only a total of 73 respondents were transgender (Mathy, 2002). The instrument utilized was a "76-item questionnaire," which expanded upon a previously existing instrument (Mathy, 2002). Suicidal ideation was determined by a simple "yes or no" response and transgender participants were offered the opportunity to declare their gender identity as "*vis-[à]-vas* male and female" (Mathy, 2002, p. 53). A "Pearson x^2" was utilized to cross-analyze the various gender identities and sexualities of the respondents (Mathy, 2002, p. 55).

Strengths. Although this article has its limitations (see "Weaknesses" below), there are some areas that will prove beneficial to the author. Utilizing this article in conjunction with the others obtained thus far further illustrates the need for further research into transgender suicide. In addition, this article adds to the author's current body of literature via the addition of alcohol and/or drug use being a potential causal factor in transgender suicidality (Mathy, 2002). This article also may help inform the author's own methodology design for obtaining and properly questioning potential participants in their own research.

Weaknesses. Although the data and findings obtained from this article are highly useful to the author, there are some key limitations that must be addressed. Mathy (2002) states in her article that this was a small study with few participants within the predominant target population. Given the lack of participants, this article cannot be generalized to the entire transgender population and thus must be utilized in conjunction with larger studies if the findings are going to be useful to the author in their research.

Conclusion

The literature analyzed thus far paints a rather grim picture. The rate of suicidality among the transgender population is alarmingly high and little is being done to actively address this issue (Goldblum, et al., 2012). Psychologists currently have a strong body of knowledge concerning suicidality and gender and sexuality in general; however, application of the currently existing knowledgebase is not producing a solution that is curbing the rate of transgender suicide. If psychologists wish to reduce the rate of transgender suicides, it is paramount that further research be done into suicidality within the transgender community in order to understand the unique factors, if any, that go into transgender individuals' choice to attempt and/or commit suicide so that they may obtain the appropriate form of help and live their lives as they truly are.

Chapter II: Literature Review

Research Question

Based on the evidence obtained thus far, the author's research question will be: "Are there unique suicidality markers present within the transgender population that are not present within the non-transgender population?" This research question will have a H_0 of "There are unique suicidality markers present within the transgender population that are not present within the non-transgender population" and a H_a of "There are not unique suicidality markers present within the transgender population are not are present within the non-transgender population." The author hypothesizes that unique causal factors will be found, based on the evidence gathered thus far.

How Does Transgender Suicide Fit into the Field of Psychology?

The fit of transgender suicide in the field of psychology is a rather complicated one. Prior to being able to fully analyze how this topic sits within psychology, the author feels that it is necessary to discuss suicidality itself as well as briefly address the concepts of gender identity, sexual orientation, and gender dysphoria and how and why being within the LGBT community may cause one to feel and/or become suicidal.

Suicide. Suicide, and the reasons for it, are numerous (APA, 2013; Berger, 2011). One of the most common reasons for suicide within the general population is the occurrence of major depressive disorder (MDD), which is a psychiatric disorder in which one often feels hopeless and worthless (APA, 2013). MDD is marked by symptomatology such as the following (APA, 2013)

- Despair

- Feelings of worthlessness

- Feelings of hopelessness

- Feelings of guilt

Within the non-LGBT population, causal factors of MDD can include things such as breakups, divorce, and other major life changes (Berger, 2011). Although life changes can be a causal factor, it is important to note that MDD can sometimes be biologically based and appears to center around low levels of serotonin, which is a neurotransmitter that is responsible for regulating one's mood (APA, 2013; Carlson, 2014).

Posttraumatic Stress Disorder. Another psychological disorder that might lead to suicidality, regardless of the population in question, is posttraumatic stress disorder (PTSD) and complex posttraumatic disorder (C-PTSD) (APA, 2013). It should be noted that C-PTSD is not found within the current edition of the DSM; however, it may be included upon release of the DSM-V-TR or DSM-VI. PTSD centers around reliving traumatization, whereas C-PTSD appears to revolve around re-experiencing prolonged traumatization (APA, 2013; Berger, 2011; Carlson, 2014). Trauma can take many different forms and does not have to be something such as experiencing war (Carlson, 2014; Berger, 2011; APA, 2013). It is entirely possible, for example, that someone that experienced abuse, whether psychological or physical, as a child will have C-PTSD (Berger, 2011). Within the LGBT population, particularly within transgender individuals, psychological abuse is extremely common and can be a key factor in occurrences of MDD and/or PTSD (Berger, 2011). This will be discussed in greater depth later in this document.

Gender Identity. On a basic level, gender identity is defined as with which gender one identifies (University of Wisconsin [UW], 2016). It is important to note that gender, which is essentially how one's mind is wired, is not limited to just male and female (UW, 2016). There are several different gender identities that serve as spectrum for gender (UW, 2016). Male and female serve as polar opposite ends of the spectrum; however, within the middle, one finds non-binary gender identities, which are neither male or female, but rather genderless or a combination of male and female (UW, 2016). The full range of the spectrum of gender identities is beyond the scope of this document; however, the author has included a self-designed chart in order to assist the reader in understanding how the various gender identities are arranged (see Figure 2).

Figure 2: Illustration of the range of gender identities

Sexual Orientation. Like gender identity, there are several different sexual orientations and a full discussion of them is once again beyond the scope of the document. However, in order to be able to understand the issues surrounding transgender suicide, it is necessary to have a working knowledge of sexual orientation. Sexual orientation is defined as to whom one is attracted, and generally takes the form of heterosexuality, bisexuality, pansexuality, and homosexuality; however, it should be noted that a multitude of other sexualities exist (UW, 2016; Berger, 2011; APA, 2007).

Heterosexuality is defined as attraction to the opposite gender exclusively, bisexuality is defined as attraction to one's own gender identity and one other gender, pansexuality is defined as being attracted to all genders with a disregard for one's sex, and homosexuality is defined as being attracted to one's own gender exclusively (UW, 2016; APA, 2007). Although sexual orientation and gender identity are two distinct things, the psychological stress they can cause often comes in tandem and can be an issue when considering transgender and LGBT MDD and PTSD occurrences (APA, 2013; Berger, 2011; Shafter & Kipp, 2014). The author has included a self-designed table below in order to aide in defining the various sexual orientations. More comprehensive definitions can be found in Appendix B.

Table 2: A brief overview of the various sexual orientations

Sexuality	Definition
Heterosexuality	Attraction to one's opposite gender (UW, 2016)
Bisexuality	Attraction to one's own gender and the opposite gender (UW, 2016)
Pansexuality	Attraction to all genders, regardless of sex (UW, 2016).

	This may also be thought of as being attracted to personality (UW, 2016).
Homosexuality	Attraction to one's own gender (UW, 2016)

Gender Dysphoria. By definition gender dysphoria is a psychological condition that "refers to an individual's affective/cognitive discontent with [their] assigned gender" (APA, 2013, p. 451). The creation of gender dysphoria was not meant to pathologize transgender individuals, but rather to give them an avenue in which to receive medical treatment in order to live as the gender they truly are (APA, 2013). The disorder is marked by a number of symptoms, which differ for children/adolescents and adults (APA, 2013). In children, some of the symptoms can include preferring play activities that are normally associated with the opposite gender, a strong dislike or outright rejection of gender standards, and "a strong dislike of one's sexual anatomy" (APA, 2013, p. 452). Some examples of this can include boys preferring to play with dolls or "dress up" and not wanting to engage in violent play or other typically masculine activities (APA, 2013; Berger, 2011).

In adults, gender dysphoria symptoms are generally marked by wanting to be the opposite gender and wanting to have the opposite genitalia of what was assigned at birth (APA, 2013). Despite the differentiation in symptomatology, the treatment for gender dysphoria is always transitioning, which is a process in which one moves from their gender assigned at birth to their correct gender (APA, 2013; Berger, 2011; Goldblum, et al., 2012). Typically, the transition process involves the utilization of hormone replacement therapy (HRT) and gender reassignment surgery (GRS) (Goldblum, et al., 2012; Mathy, 2002). In males, HRT is accomplished by taking medications that halt the production of testosterone and then introduce estrogen in order to obtain female sex and gender characteristics; in females, this process is reversed (Goldblum, et al., 2016; Carlson, 2014). The details of GRS are beyond the scope of this document, but the procedures typically involve surgical removal and inversion of the genitalia that was assigned at birth so that either a penis and testes or a vagina can be formed (Goldblum, et al., 2016; Carlson, 2014).

Halting the Melancholic Tintinnabulation of the Church Tower Bells

The reasons that LGBT youth, particularly transgender youth, may become susceptible to MDD, PTSD, and suicidality is due to homophobia, which is fear and discrimination directed at LGBT individuals, at school and at home (Poteat, Mereish, Koenig, & DiGiovanni, 2011). Poteat, Mereish, Koenig, and DiGiovanni (2011) found that although homophobia in schools and at home played a large role in LGBT youth developing psychopathologies such as MDD and suicidality, evidence suggests that the largest issue is a lack of parental understanding and support. What this may indicate is that a lack of parental acceptance of one's LGBT status plays a large role in one's choice to commit suicide. In addition, lack of parental support had pejorative effects on LGBT youth's academic performance, particularly when homophobic victimization was in conjunction with a lack of parental support (Poteat, Mereish, Koenig, & DiGiovanni, 2011). An additional study from the United Kingdom builds upon these findings and notes that current evidence suggests that suicide and self-harm may be coping mechanisms that LGBT, particularly transgender, individuals are utilizing as a way to cope with the homophobia that they are experiencing (Scourfield, Roen, & McDermott, 2008). In addition, recent findings further compound the issue of transgender suicide by finding that in addition to the aforementioned effects of bullying, discrimination, and poor parental support, transgender people are put at further risk of suicide via sexually-based violence, which further raises their risk of engaging in self-harm and/or suicidal behavior (Haas, Rodgers, & Herman, 2014).

Synthesis of Research Findings

Overall, the research presented thus far continues to indicate a grim situation. All the studies that were analyzed, particularly Haas, Rodgers, and Herman (2014) and Scourfield, Roen, and McDermott (2008), indicate that transgender suicides are occurring at an alarmingly high rate due to various potential reasons, such as bullying, discrimination, and a lack of parental support.

Additionally, the APA (2013) acknowledges the conditions of gender dysphoria, MDD, and PTSD and their prevalence in LGBT and transgender individuals; however, evidence suggests that there is not enough of an effort being made in order to get transgender and

LGBT people accepted by society so that they may receive appropriate treatment, such as HRT and GRS (Poteat, Mereish, Koenig, & DiGiovanni, 2011).

The issue is further compounded by stories such as Leelah Alcorn's suicide, which made international news (Weems, 2015). Alcorn's death was met with public outcry; however, it was not enough to get the public and policy makers to make a greater effort to prevent situations like this from occurring in the future (Weems, 2015; Goldblum, et al., 2012). Evidence suggests that parental support and making the medically necessary tools to change one's gender, such as HRT and GRS, more easily available may help to address the issue, but further research is necessary to validate the claims (Carlson, 2014; Goldblum, et al., 2012; Berger, 2011).

As previously mentioned, parental abuse may also be a factor in transgender suicidality (Poteat, Mereish, Koenig, & DiGiovanni, 2011). It is possible that abuse may stem from homophobia; however, it is marked by several key "symptoms" (Ackerman, 2010). The predominant form of abuse that would be seen in LGBT and transgender individuals would most likely be psychological abuse (Ackerman, 2010). Psychological abuse is marked by actions as talking down to an individual, insulting them, restricting access to money, and various other malicious actions (Ackerman, 2010).

A colloquial example of talking down to an individual would be stating something such as: "You're retarded," or a similar phrase which diminishes one's intellectual capacity (Ackerman, 2010). The full scope of what constitutes psychological abuse is beyond the scope of this document; however, it is paramount to note that it may very well play a large role in transgender individuals' decisions to commit suicide (Goldblum, et al., 2012; Ackerman, 2010).

Evaluation of Studies

The studies, and one periodical, that are utilized within this literature review work in conjunction to conjure a very grim situation. Rather than indicating substantial differences, Poteat, Mereish, Koenig, and DiGiovanni (2011), Scourfield, Roen, and McDermott (2008), and Hass, Rodgers, and Herman (2014) all illustrate the fact that there are extensive issues occurring within the transgender community and its psychological health that are leading to large numbers of suicides. Poteat, Mereish, Koenig, and Digiovanni (2011) and Scourfield, Roen, and McDermott (2008) highlight evidence that suggests that homophobia found within the school system, such as via bullying, and a lack of parental acceptance of individuals' transgender status is contributing to a large amount of the transgender community's large number of MDD and PTSD cases. In addition, Ackerman (2010) and the APA (2013) highlight that parental abuse may also play a contributing role in transgender individuals' suicidality since the home is supposed to be a place of reprieve and acceptance (Berger, 2011). These issues are analyzed further in the "Research Gaps" and "Contribution of Existing Research" sections below.

Research Gaps

Despite the large body of research that encompasses LGBT individuals, transgender individuals, and suicidality, there is an apparent lack of suicidality research within the transgender community itself. Poteat, Mereish, Koenig, and Digiovanni (2011) and Scourfield, Roen, and McDermott (2008) highlight that issues such as poor parental acceptance of being LGBT and/or transgender can increase one's risk of suicidality, there is not any current research on the suicidality markers that may be unique to transgender individuals.

As previously stated, the APA (2013) outlines a list of symptomatology for which to be aware when working with patients that might be suicidal; however, this list is generalized to a standard population rather than minority groups, which may or may not differ in their presentation of suicidality markers. For example, such a distinction can be found when examining suicidality within adolescents and adults (Berger, 2011). Although there may or may not be such a distinction in the suicidality markers of transgender individuals, the current evidence suggests that if further research is not conducted, psychologists and social institutions, such as schools and one's family, will not be able to do much to curb the LGBT and transgender suicide epidemics (Goldblum, et al., 2012).

Contribution of Existing Research

The current body of literature has offered a large contribution to understanding and treating psychopathologies such as PTSD and MDD via methodologies such as cognitive behavioral therapy (CBT) and psychopharmacological substances such as selective serotonin reuptake inhibitors (SSRIs) as well as the disorders' biological causal factors; however, it should be noted that further research into psychopathologies' biological causal factors is a constantly evolving facet of psychology (Preston & Johnson, 2016; Carlson, 2014). In addition, the literature has given us a fairly solid understanding of LGBT issues such as suicidality and struggling with one's gender identity and how one may work towards coming to accept and live as their gender identity, in the case of incongruence (Mathy, 2002; Goldblum, et al., 2012; Hass, Rodgers, & Herman, 2014).

In addition, the existing body of research has also given us the interpersonal-psychological theory of suicide (IPTS), which essentially states that suicidal ideation is based on three factors that influence each other (Davis, Witte, & Weathers, 2014). The three factors are: (1) "perceived burdensomeness," (2) "thwarted belongingness," and (3) "acquired capability" (Davis, Witte, & Weathers, 2014, p. 611). The author has constructed a diagram (see Figure 3), which may assist in comprehending IPTS.

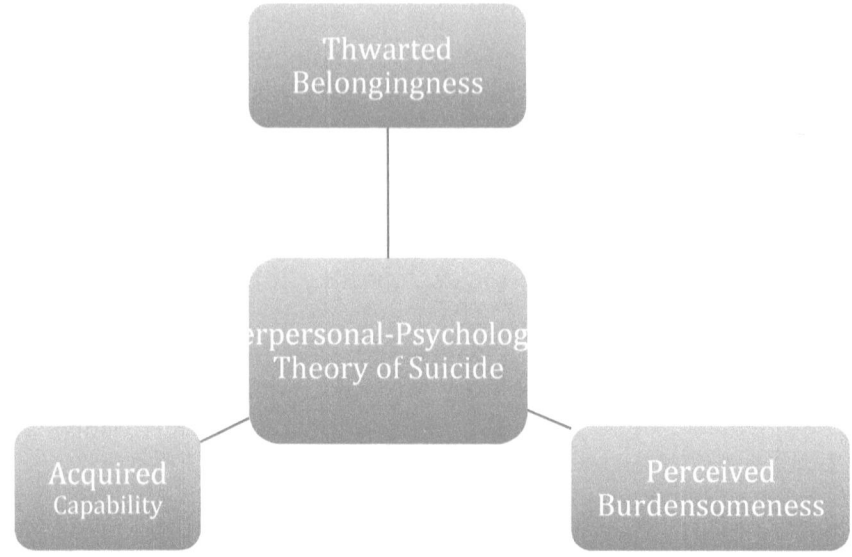

Figure 3: Illustration of IPTS based on Davis, Witte, and Weathers (2014)

As illustrated in Figure 3, each of the three factors influence the individual's suicidality; thus, if one perceives oneself as being a burden, then that person is at risk of committing suicide (Davis, Witte, & Weathers, 2014). If any of the other two factors become present, then the individual's risk of suicide increases further (Davis, Witte, & Weathers, 2014). Unfortunately, how the suicidality markers of IPTS would be applied to and exhibit themselves within the transgender population is currently unknown.

Conclusion

Examination of the literature as a whole is, at best, depressing. Evidence from the vast majority of the literature appears to point to aspects such as homophobia and/or transphobia, which is fear and/or hatred of transgender individuals, from transgender people's parents, family members, teachers, and classmates as a large aspect in transgender individuals' choice to commit suicide (Goldblum, et al., 2012; Mathy, 2002; Poteat, Mereish, Koenig, & DiGiovanni, 2011;

Scourfield, Roen, & McDermott, 2008; UW, 2016). Additional evidence suggests that parental psychological abuse due to LGBT and transgender individuals' sexual orientations and gender identities may also be a contributing factor as their lack of acceptance may play a role in IPST (Ackerman, 2010; Davis, Witte, & Weathers, 2014).

As outlined above, further evidence suggests that IPST may be serving as a "foundation" for transgender individuals' suicidal ideation. If a transgender individual feels that they are not accepted by their family, for example, they may feel that they are not wanted, which would cause them to fall within the category of thwarted belongingness (Davis, Witte, & Weathers, 2014). In addition, if their parents make the individual feel that transition is too costly, that may contribute to perceived burdensomeness, which would in turn further compound the risk of suicide (Davis, Witte, & Weathers, 2014).

Although the evidence appears to lay a strong framework as to the "why" behind transgender suicide, psychologists still do not know if being transgender leads to unique suicidality markers. Based on the evidence in Davis, Witte, and Weathers (2014) and Goldblum, et al. (2012), IPTS and the reasons behind transgender individuals' reasons for committing suicide might be co-morbid. That is to say, if IPTS factors are present, it may possible that there is crossover among the expressed suicidality symptomatology and the underlying reasons for the individuals' suicidality.

Additional study and evidence will be necessary to confirm or deny the author's hypothesis; however, the evidence makes it clear that extensive study is necessary if psychologists are going to be able to understand and treat the potentially unique symptomatology of suicidality within the transgender population.

Chapter III: Methodology, Design, and Data

Research Method(s)

For the purpose of this research, the author will utilize a self-designed survey that will include a mixture of multiple choice responses, Likert-type scales, and subject response questions, as outlined in Cozby and Bates (2012). Even though statistical analysis will be applied to this survey's results, this type of research is predominantly qualitative in nature as the author is working with a survey and its responses rather than more "numerically-based" data, such as the mean time it took a psychotropic medication to work or data from a neuroscientific experiment (Cozby & Bates, 2012).

Support of Method(s)

Due to the nature of the subject matter at hand and the population being studied, qualitative survey research is the best approach for this study. The subject matter is something that is deeply personal to the participants, thus it is best to engage them with a survey rather than conducting an interview directly. Conduction of an interview would potentially "contaminate" the data as there is the potential for subjects to not feel comfortable speaking with the author, thus they may give incorrect data, which would cause issues with the final data. In addition, there is further justification for survey research as other approaches, such as requesting subjects' psychiatric records is unethical as well as unjustifiable in this case (Cozby & Bates, 2012).

In addition, the author wishes to respect the subjects' histories as much as possible. This is a topic that is incredibly personal to the subjects, thus they need to be treated with upmost respect. The author feels that the best way to do this is to give the participants a survey so that they feel their privacy is completely intact. In addition, they would most likely feel less comfortable speaking to the author directly due to the proximity of the author's place of residence to the research site. Hence, from an ethical and practicality standpoint, survey research is the least "invasive" method that is available to the author at this time (Cozby & Bates, 2012).

Strengths and Weaknesses

Strengths. The predominant strengths of this design are found within the main advantages of self-designed survey research and the study's ability to reach the target population.

Weaknesses. There are two predominant weaknesses in this design that must be addressed: (1) the limitations of survey research itself and (2) the issues that surround a self-designed survey rather than a pre-established one. Each of these facets will be discussed below individually. Weaknesses of the study itself may be found under "Limitations."

Limitations of Survey Research. Survey research comes with the issues of potential bias from both the researcher and the responder (Cozby & Bates, 2012). Essentially, there is the potential that the researcher's personal biases might come forth when the researcher is going through responses and/or coding "self-answer" questions (Cozby & Bates, 2012). Additionally, responder bias may occur if the responders feel that there is a need to "impress" the researcher, in fear of giving a "wrong answer," for example (Cozby & Bates, 2012). These issues are discussed in greater depth in "Limitations."

Issues with Self-Designed Surveys. Self-designed surveys come with their own set of potential issues. The predominant problem here is with the potential reliability and validity of the questions themselves (Cozby & Bates, 2012). For example, if a researcher elects to use a pre-established set of questions, the validity and reliability of that psychometric instrument is already known; however, if a researcher develops their own set of questions it will be unknown going into the research if, for example, these sets of questions are biased towards female subjects (Cozby & Bates, 2012). More information concerning these issues may be found under "Limitations."

Data-Collection Tools

The predominant instruments that will be utilized to collect data for this project are Google Forms and SPSS for statistical analysis. Google Forms is a program that is provided for free by Google that permits for the creation and analysis of surveys (Google, 2016). Taking into consideration that Google is a well-known international company, Google Forms' reliability and validity will be ensured in lieu of utilizing a program that is less known (Google, 2016). The survey itself will be self-designed; however, it will utilize a combination of questions constructed by the author as well as questions from well-known psychometric instruments, such as Beck's Depression Inventory and other assessments of depression and/or suicidality.

Population and Sampling Procedures

The subjects for this study will be acquired from the University of Mary Washington's (UMW) People for the Representation of Sexual Minorities (PRISM) organization. UMW is located in Fredericksburg, Virginia, which is just outside of Washington, D.C (UMW PRISM, 2015). PRISM is a student-run organization for LGBT students and their allies to gather together and discuss LGBT-related issues (UMW PRISM, 2015). PRISM consists of a sufficient number of students $(N \cong 30)$ with approximately half of those students $(n \cong 15)$ being within the transgender population (UMW PRISM, 2015). The specific numbers of male-to-female (MTF), non-binary, and female-to-male (FTM) students is unknown.

Sampling will be done via the distribution of an electronic survey to the PRISM Facebook page so that the entirety of the target population will be reached and made aware of the survey in question. Due to the fact that the author will not be aware of whom is transgender upon initial release of the survey, true sampling will occur once all data have been collected so that the author may go through and code responses, if needed. This survey and the author's sampling procedures are discussed in greater detail in "Data Collection" and "Data-Analysis Process."

Data Collection

The process of data collection will be as straightforward as possible. First, approval for experimentation on human subjects will be obtained from the institutional review boards (IRB) of UMW and Southern New Hampshire University (SNHU). Upon obtaining approval from UMW's and SNHU's IRBs, the author will proceed with distribution of the survey, which will be administered to subjects via email to PRISM's members.

Informed consent will be obtained on the first screen of the survey (see Appendix C for a sample informed consent statement). Once the author has obtained the proper consent, the subjects will take the author's survey. Upon collection of all data, the author will begin the data analysis process, which is outlined below. In order to facilitate subjects' participation in the author's research, the author will conduct a random drawing of all participants for a $50 Starbucks and/or Blackstone gift card.

Ethics. The ethics of this project are discussed in greater depth in Milestone IV; however, there are some key points that must be noted here. The ethical practice of clinical psychology, the teaching of psychology, and psychological research is predominantly governed by the American Psychological Association's *Ethical Principles of Psychologists and Code of Conduct*, which is often shortened to the *Ethics Code* or *Code of Ethics* (American Psychological Association, 2010).

The Ethics Code is incredibly comprehensive; however, for the purposes of this research, the researcher will predominantly be dealing with Sections 3, 4 and 8 (American Psychological Association, 2010). Section 3 deals with "human relations," Section 4 outlines the practices for "privacy and confidentiality," and Section 8 outlines research practices (American Psychological Association, 2010). Each of these sections are further divided into subsections, such as Sections 3.04 and 3.10, which outline the practices of "avoiding harm" and "[obtaining] informed consent" prior to engaging in research (American Psychological Association, 2010). Although Section 4 is predominantly clinical in nature, it outlines the need to maintain confidentiality, which is highly applicable to this project (American Psychological Association, 2010). In addition, Section 8's practices for research will be used almost in its entirety, thus discussing it in depth is beyond the scope of this document (American Psychological Association, 2010). The author notes here that the "highlights" from Section 8 include the various ways in which informed consent must be obtained for research, the need to utilize the IRB, and the need to debrief subjects either prior to research or immediately after the conduction of research (American Psychological Association, 2010).

This project deals with subjects that are highly personal to the participants, thus by engaging in such research, it is highly likely that they may be triggered (see Appendix B for a definition of this term) by engaging in the author's research. In order to combat this possibility, the author has ensured that all subjects have the option to stop the research at any time in the event that psychopathological symptomatology arises during the research process. In addition, the author will guarantee to the subjects that no personal data will be collected, thus ensuring that their identities remain anonymous.

Data-Analysis Process

Upon collection of the data, the data will be analyzed using a predominantly statistical approach, assuming that no coding of responses is required. If it is necessary to code responses, the author will properly code the subjects' survey responses prior to engaging in statistical and qualitative analysis. All statistical calculations will be conducted utilizing SPSS in addition to any statistical tools that are included within Google Forms.

All data will first be transferred from Google Forms to SPSS. Upon importation of the data and any necessary coding of responses, statistical analysis will be conducted. The researcher will use mean, median, and mode as well as a multivariable analysis of variance (MANOVA), specifically Wilk's Lamda, and logistic regression calculation in order to compute the relationships, if any, between suicidal markers and one's gender identity. In addition, the measures of central tendency will be utilized to determine the mean levels of suicidality markers for each subgroup of transgender individuals. Upon completion of all computations, the data will be transferred to the "Results" section to be converted to properly displayed data for readers' analysis.

Although the calculations will be conducted utilizing SPSS and Google Forms, the author has elected to incorporate the equations that will be utilized in order to facilitate accurate reproduction of this study. The equation for Wilk's Lamda may be found below:

Equation 1: Wilk's Lambda (Field, 2013, p. 641)

$$\Lambda = \prod_{i=1}^{s} \frac{1}{1 + \lambda_i}$$

The equation for logistic regression may be found below:

Equation 2: Logistic Regression Equation (Field, 2013, p. 341)

$$VIF = \frac{\sum_{i=1}^{k} VIF_i}{k}$$

A brief discussion of each calculation's appropriateness to the author's research may be found in "Justification of Statistical and/or Qualitative Analysis."

Justification of Statistical and/or Qualitative Analysis

This research will utilize a mixture of both statistical and qualitative analysis, depending upon the author's findings. If the author selects only multiple choice questions for the survey, then the research will entirely statistical as the author will only be analyzing the subjects' responses in a statistical manner, such as with Wilk's Lambda (see Equation 1) and logistic regression (see Equation 2). However, if the author chooses to utilize a "free response" type of question, then qualitative analysis will become necessary as the author will need to code the subjects' answers.

If qualitative analysis is utilized, the author will ensure reliability and validity by coding the responses into set categories that are already "pre-established" within the transgender and LGBT community. For example, if a subject indicated that they are a "demi-girl," then the author would code that response as falling within the non-binary category. By utilizing such pre-established categories, it is a lot more likely that the data will be kept reliable and valid for statistical analysis (Cozby & Bates, 2012).

Limitations and Assumptions

Limitations. Unfortunately, there are some key limitations to this study that must be acknowledged. Firstly, the potential population size $(N \cong 30)$ is fairly small by statistical standards, thus generalizing the results to the entire transgender and LGBT populations would not be feasible. For proper generalizability, one would need a population of roughly $N \geq 300$. In addition, a subpopulation of $n \cong 15$ greatly limits the possibility that there will be a sufficient sample size in order to assess unique suicidality factors since not all participants will have expressed suicidal ideation.

In addition, survey research comes with inherit limitations that should be acknowledged (Cozby & Bates, 2012). With survey research, one runs the risk of both interpreter bias and responder bias (Cozby & Bates, 2012). The way this survey has been designed (see Appendix C for sample questions) opens up the possibility of both types of biases as there is a chance that participants will respond with their own answer if there is an "other" option and the participants fill in their own answer (Cozby & Bates, 2012). Participants may put in answers that they think the researcher wants to see; on the other hand, if the researcher needs to code the responses, there is the possibility that the researcher will be biased in the way each response is coded for statistical analysis (Cozby & Bates, 2012). It should also be noted that the survey is self-designed, thus validity and reliability will not be as strong in comparison to a pre-established survey or psychometric tool.

Assumptions. This study comes with some assumptions, which partially coincide with the limitations that were discussed above. The author is operating under the assumption that a sufficient number of subjects will be available in order to obtain statistical and clinically significant data. In addition, there is an assumption that there will a sufficient number of subjects that are/were suicidal so that their presentation of suicidality can be appropriately analyzed. As the author indirectly acknowledged in "Limitations," it is possible that none of the transgender members of PRISM will have expressed or experienced any degree of suicidality whatsoever, in which case the study itself would have to be somewhat redirected when the findings and data are discussed.

Dissemination of Findings

There are multiple viable avenues that may be utilized in order to ensure the proper dissemination of the author's findings to individuals that are interested in the research. Rather than selecting one possibility of dissemination, the author has elected to analyze each major possibility briefly below.

PowerPoint Presentation. One of the simplest possibilities would be to conduct a PowerPoint presentation to all interested parties. Within this presentation, all data will be summarized and explained in such a way that it is easily comprehensible by individuals that are psychologists, "non-scientists," and/or the LGBT population. This methodology would not reach many individuals; however, it would ensure that all interested individuals are made aware of the author's research.

Publication in a Journal. Publication of the author's findings in a reputable academic journal is the least likely possibility for dissemination of the findings due to the sheer difficulty of getting scholarly literature published. However, this avenue would reach all interested parties, particularly psychologists and psychology students, as well interested "lay people."

Presentation of Findings at a Conference. Presentation of the author's findings at a scholarly conference would most likely be the most feasible option for dissemination of the author's findings. By presenting a conference, the data will be disseminated to a wide audience, which would most likely open the door to additional utilization of the data and research beyond the author's original intentions.

Chapter IV: Conclusion

The sum of this research paints a dark and pejorative picture at best. The field of contemporary psychology science has been presented with something which is does not have the current body of research to properly address (Goldblum, et al., 2012; Mathy, 2002). However, research is beginning to point to possible causes of the high degree of transgender suicide (Goldblum, et al., 2012).

On the one hand, there is the possibility that physical and/or psychological abuse of transgender individuals is causing conditions such as PTSD and MDD, which in turn leads to one's desire to commit suicide because of the trauma (American Psychiatric Association, 2013; Berger 2011). An additional possibility may be that transgender individuals that are rejected by their family and peers, and thus cannot transition are left at a "crossroads" in which they feel they have no other possible options (Scourfield, Roen, & McDermott, 2008). In addition, one must consider the possible involvement of IPTS in one's decision to commit suicide, which essentially states that the causal factors of suicide build upon each other, as outlined in Figure 3.

Although psychological science has made a large degree of progress in the field of gender and sexuality, several pressing clinical issues still remain:

- Do transgender individuals present unique suicidality

  markers?

- If these unique markers are present, how should they be

  handled clinically during treatment?

- Is there anything that can be done to meet the unique needs of

  transgender individuals that is not being done now that can

  help to stop their suicidality?

Psychological science will not be able to address issues such as this without further research into gender and sexuality and the transgender community as a whole. Based on the research, the author believes that additional research into IPTS and how it may be applicable to the transgender community would be beneficial to the overall body of literature. Although the proposed same size is small, this study would be a possible step into the right direction towards better understanding transgender suicide and how it may be prevented in the future so that additional lives are not unnecessarily lost.

Chapter V: Professional Reflection

Overall Experience and Issues

The author's experience with the capstone project was both positive and pejorative. On the positive end of the spectrum, the author found the experience of the actual writing and research to be fairly easy, which was surprising given the length of the final document that was produced in contrast to previously completed work. The only true pejorative aspect of the project was attempting to figure out the best statistical test to use as the author is not as well versed in statistics as they would like; however, Dr. MacCarty was a wonderful asset in that regard, so the issue was rapidly negated. The only other issue that the author experienced was attempting to find sufficient time to work on the project as various issues, such as a close friend/roommate's physical and psychological health issues acting up, arose throughout the semester; however, Dr. MacCarty and Danielle, the author's advisor, were incredibly accommodating.

Giving and Receiving Feedback from Peers and Dr. MacCarty

The experience of giving and receiving feedback from peers and Dr. MacCarty was nothing new for the author as the author came into the MS program with a BA in English/Creative Writing and an AAS in education, thus the giving and taking of feedback was an essential part of the author's previous programs. The author found almost all criticisms and suggestions to be incredibly helpful; however, the author found difficulty in not giving "English major style" feedback to their classmates, which more than likely would have been interpreted as being "overly harsh." Generally speaking, the feedback and suggestions were helpful and were incorporated as the need arose.

What Led to This Project?

The background and of this project is mixed and came from both the MS program and the author's alma matar, the University of Mary

Washington (UMW). While the author attended UMW, the author was exposed to the University's very strong and active Lesbian, Gay, Bisexual, Transgender, and Questioning (LGBTQ) community. The author quickly befriended many of these individuals, who took it upon themselves to tell the author their stories of coming out, their experiences with gender dysphoria, and the process of transitioning. By the time the author graduated from UMW, the author had a strong research interest in clinical psychology and human sexuality and gender. These interests were then carried over into the MS program and fueled the vast majority of the author's research projects whenever feasible, which in turn led to author furthering their expertise in these areas.

Taking This Project Into the Workplace and a Doctoral Program

Post-graduation, the author is planning to accept an adjunct psychology professorship at Germanna Community College (GCC) in Fredericksburg, Virginia, which is where the author's educational journey began. In addition, the author is planning to pursue a Doctor of Psychology degree in Clinical Psychology, beginning in March of 2017.

GCC

Currently, GCC does not offer any psychology courses that deal with human sexuality and gender, outside of what is covered in their Introduction to Psychology course. One way the author is planning to use this capstone project and their MS, outside of teaching psychology, is to potentially propose an introductory course on gender and sexuality, which would expand GCC's course offerings and put the College on par with UMW's course offerings in psychology. In addition, the author strongly believes that everyone should have at least some sort of background in psychology, even if it is just an introductory course as being able to understand the behavior of others, even to a small degree, can be key in helping to reduce acts of violence and prejudice.

Doctoral Program

If all goes well, the author is planning to expand upon this capstone by examining the possible origins of gender identity. That is to say, the author is going to build upon transgender individuals' high rates of suicidality and examine whether or not there is a biological and/or developmental origin to being transgender. This in turn would, hypothetically, advance psychological science, the care of transgender individuals, and help further address the issue of transgender individuals being told that being transgender is a "choice" and/or a "lifestyle." Upon completion of the author's doctorate, the author is hoping to open a small private practice which will focus on disorders such as anxiety, PTSD, and LGBTQ/transgender issues.

Conclusion

The experience of moving through this MS program and writing this capstone project has been nothing short of surreal. It is something that the author never thought that they would be able to accomplish, let alone strongly enough to be able to pursue a doctoral degree. Despite the fact that this experience is drawing to a close, the author looks forward to seeing what doors will open as they continue to finish their education, and hopefully help to make the world a better place utilizing psychological science and psychological education.

Appendix A: List of Acronyms

American Psychiatric Association (APA)
Cognitive Behavioral Therapy (CBT)
Complex Posttraumatic Disorder (C-PTSD)
Diagnostic and Statistical Manual of Mental Disorders-V (DSM-V)
Female-to-Male (FTM)
Gender based victimization (GBV)
Gender Reassignment Surgery (GRS)
Hormone Replacement Therapy (HRT)
Institutional Review Board (IRB)
Interpersonal-Psychology Theory of Suicide (IPTS)
Lesbian, Gay, Bisexual, and Transgender (LGBT)
Major Depressive Disorder (MDD)
Male-to-Female (MTF)
Posttraumatic Stress Disorder (PTSD)
Selective Serotonin Reuptake Inhibitors (SSRIs)
Southern New Hampshire University (SNHU)
University of Mary Washington (UMW)

Appendix B: Glossary

Bisexual: An individual that is attracted to their own gender and another

Gender Identity: An individual's internal sense of being male, female, both, neither, or something else. Since gender identity is internal, one's gender identity is not necessarily visible to others

Gender Reassignment Surgery (GRS): GRS is a surgical procedure that is utilized to assist transgender individuals in obtaining the genitalia that is consistent with their gender identity

Gender: A social combination of identity, expression, and social elements related to masculinity and femininity. Includes gender identity (self-identification), gender expression (self-expression), social gender (social expectations), gender roles (socialized actions), and gender attribution (social perception)

Heterosexual: A person emotionally, physically, and/or sexually attracted to people of different sex or gender

Homophobia: Fear and discrimination against LGBT individuals

Hormone Replacement Therapy (HRT): A medical procedure that is utilized with transgender individuals in order to assist them in gaining sexual and gender chracteristics of their gender identity

Interpersonal-Psychology Theory of Suicide: A theory found within clinical psychology that bases one's risk of suicidality on three specific markers that influence each other

Lesbian: A female-identifying individual that is only attracted to other female-identifying individuals

Non-Binary: 1) Describes a gender identity that is neither female nor male; 2) Gender identities that are outside of or beyond two traditional concepts of male or female

Nontransitioning Trans* People: Individuals that identify as transgender, but elect not to transition for various personal reasons

Pansexuality: Attraction to all genders, regardless of one's biological sex

Sexual Orientation: An individual's physical and/or emotional attraction to and desire to sexually or emotionally partner with specific genders and/or sexes

Suicidal Ideation: Suicidal thoughts or a preoccupation with suicide

Suicidality: One's risk of suicide

Transgender: An umbrella term describing a diverse community of people whose gender identity differs from that which they were designated at birth; 2) Expressions and identities that challenge the binary male/female gender system in a given culture; 3) Anyone who transcends the conventional definitions of man and woman and whose self-identification or expression challenges traditional notions of male and female

Transman: An individual that was born female, but transitioned to male

Transphobia: Fear and/or hatred of transgender individuals

Transwoman: An individual that was born male, but transitioned to female

Triggered: This is a term that refers to one's clinical psychopathology symptomatology being engaged by a stimulus and/or stimuli

References

Ackerman, M. J. (2010). *Essentials of forensic psychological assessment* (2nd ed.). Hoboken, NJ: John Wiley & Sons, Inc. .

American Psychiatric Assocation. (2013). *Diagnostic and statistical manual of mental disorders* (5 ed.). Washington, D.C.: American Psychiatric Publishing.

American Psychological Association. (2002). *Glossary of psychological terms.* Retrieved October 23, 2016, from apa.org: http://www.apa.org/research/action/glossary.aspx?tab=19

American Psychological Association. (2007). *APA dictionary of psychology.* (G. R. VandenBos, Ed.) Washington, D.C.: American Psychological Association.

American Psychological Association. (2010, June 1). *Ethical principles of psychologists and code of conduct including 2010 amendments.* Retrieved November 27, 2016, from apa.org: http://www.apa.org/ethics/code/

Berger, K. S. (2011). *The developing person through the life span* (8th ed.). New York, NY: Worth Publishing .

Carlson, N. R. (2014). *Foundations of behavioral neuroscience* (9th ed.). Boston: Pearson.

Cozby, P. C., & Bates, S. C. (2012). *Methods in behavioral research* (11th ed.). New York, NY: McGraw Hill.

Davis, M. T., Witte, T. K., & Weathers, F. W. (2014). Posttraumatic stress disorder and suicidal ideation: The role of specific symptoms within the framework of the interpersonal-psychological theory of suicide. *Psychological Trauma: Theory, Research, Practice, and Policy, 6*(6), 610 - 618.

Field, A. (2013). *Discovering statistics using IMB SPSS statistics and sex and drugs and rock 'n' roll* (4th ed.). Los Angeles, CA: SAGE .

Goldblum, P., Testa, R. J., Pflum, S., Hendricks, M. L., Bradford, J., & Bongar, B. (2012, October). The relationship between gender-based victimization and suicide attempts in transgender people. *Professional Psychology: Research and Practice, 43*(5), 468 - 475.

Google. (2016). *Forms*. Retrieved November 20, 2016, from Google.com: https://www.google.com/forms/about/

Haas, A. P., Rodgers, P. L., & Herman, J. L. (2014, January). *Suicide attempts among transgender and gender non-conforming adults: Findings of the national transgender discrimination survey.* Retrieved October 2, 2016, from williamsinstitute.law.ucla.edu: http://williamsinstitute.law.ucla.edu/wp-content/uploads/AFSP-Williams-Suicide-Report-Final.pdf

Mathy, R. M. (2002, December 1). Transgender identity and suicidality in a nonclinical sample: Sexual orientation, psychiatric history, and compulsive behaviors. *Journal of Psychology & Human Sexuality, 14*(4), 47 - 65.

Poteat, V. P., Mereish, E. H., Koenig, B. W., & DiGiovanni, C. D. (2011). The effects of general and homophobic victimization on adolescents' psychosocial and educational concerns: The importance of intersecting identities and parent support. *Journal of Counseling Psychology, 58*(4), 597 - 609.

Preston, J., & Johnson, J. (2016). *Clinical psychopharmacology made ridiculously simple* (8th ed.). Miami , Florida: MedMaster, Inc. .

www.ingramcontent.com/pod-product-compliance
Lightning Source LLC
Chambersburg PA
CBHW021928170526
45157CB00005B/2231